SCOTTISH STEAM 1948-1966

The History Press books by the same author:

Steam in Scotland: The Railway Photographs of R.J. (Ron) Buckley

Southern Steam: The Railway Photographs of R.J. (Ron) Buckley

Steam in the North East: The Railway Photographs of R.J. (Ron) Buckley

Great Western Steam: The Railway Photographs of R.J. (Ron) Buckley

Steam in the East Midlands and East Anglia: The Railway Photographs of R.J. (Ron) Buckley

London Midland Steam: The Railway Photographs of R.J. (Ron) Buckley

Early and First Generation Green Diesels in Photographs

Steam in the 1950s: The Railway Photographs of Robert Butterfield

Diesel Power in the North Eastern Region

Scottish Steam 1948–1967

SCOTTISH STEAM 1948-1966

The Railway Photographs of Andrew Grant Forsyth

BRIAN J. DICKSON

The History Press

Front Cover: Monday, 1 August 1955. The footplate crew of ex-NBR Class J (LNER Class D30) 'Superheated Scott' No. 62426 *Cuddie Headrigg* are seen in discussion with the train guard prior to departure from Stirling Station at the head of a Fife-bound 'stopper'.

Back Cover: Tuesday, 12 August 1952. At Maud Station Class B1 4-6-0 No. 61324 is about to depart at the head of a passenger working from Peterhead to Aberdeen. Constructed by the NBL during 1948, she was allocated to Kittybrewester shed in Aberdeen and would be withdrawn in 1965.

Half-title Page: Friday, 15 August 1952. Class D41 4-4-0 No. 62248 waits to depart from Craigellachie Station at the head of a working to Boat of Garten via the Speyside line.

Title Page: Monday, 11 August 1952. Class B12 4-6-0 No. 61532 has completed shunting in Ballater goods yard and is ready to depart with its train to Aberdeen.

First published 2021

The History Press
97 St George's Place,
Cheltenham GL50 3QB
www.thehistorypress.co.uk

British Library Cataloguing in Publication Data.
A catalogue record for this book is available from the British Library.

ISBN 978 0 7509 9665 5

Typesetting and origination by The History Press
Printed in Turkey by Imak

INTRODUCTION

Andrew Grant Forsyth was born in Barnet, North London, in 1923, and lived and worked in the same area all his life. A lifelong railway enthusiast, he was very much a follower of all things associated with the London and North Eastern Railway, and the greater part of his photographic collection contains images relating to that company and its pre-grouping constituents. His railway photography started in 1947 with a Kodak folding camera and an Agfa Record camera using 2¼in x 3½in (6cm x 9cm) film. He moved to 35mm film in 1950 with a Kodak Retina camera, which was followed by a Leica, a Pentax S3 and finally a Nikon F90.

His catalogue of photographs details his visits to Scotland every year from 1948 until 1958 and again from 1962 until 1966, showing the changing locomotive scene throughout that country after the nationalisation of the railways in 1948. Illustrated are the graceful lines of the ex-Great North of Scotland Railway 4-4-0s, the ex-North British Railway 'Glen' and 'Scott' 4-4-0s and numerous 0-6-0 tender and tank locomotives remaining from both these companies and the Caledonian Railway. Also reproduced are many examples of the London and North Eastern Railway express locomotive fleet together with locomotives of the former London Midland and Scottish Railway and examples of the post-nationalisation Standard locomotives of British Railways.

The gap in Andrew's catalogue of photographs from 1958 until 1962 coincides with his call-up for National Service.

In his spare time Andrew was one of the organisers of The Seafield Railway Club, which arranged visits to sheds and railway sites throughout eastern England and Scotland; he obtained the shed permits and organised train times for the visits. The club also published a regular magazine titled *The Locomotive Post* containing observations from contributing enthusiasts and detailing locomotive stock changes and movements that were taking place throughout the industry.

Prior to his death at the age of 83, his photographic collection had already been carefully managed by Initial Photographics, from where a complete catalogue can be obtained by writing to 25 The Limes, Stony Stratford, MK11 1ET.

Thursday, 19 August 1948. Standing in Inverurie Works yard, beautifully turned out after an overhaul, is ex-GER Class S69 (LNER Class B12) 4-6-0 No. 61529. With the demand for more powerful locomotives to handle the increasing traffic weights of the Great Eastern expresses, a total of eighty examples of the class were constructed at Stratford Works between 1911 and 1928. The example seen here entered service during 1914 numbered 1529 by the GER; she would become No. 8529 and later 1529 again with the LNER. Allocated to the Great North of Scotland (GNS) section during 1939, she would be withdrawn in 1960.

Thursday, 19 August 1948. Having been withdrawn from traffic during December of the previous year, ex-NBR Class M (LNER Class D31) 4-4-0 No. 2073 is still standing in the scrap line at Inverurie Works. Constructed at Cowlairs Works in 1899 and numbered 769 by the NBR, she would become No. 9769 and later 2073 with the LNER. Transferred to the GNS section during 1944, she had previously been allocated to Dunfermline and would end her days based at Kittybrewster in Aberdeen.

Thursday, 19 August 1948. Originally designed by James Johnson and introduced during 1893, the GNoSR Classes S and T were the general workhorses of that railway. The six examples of Class S entered service from Neilson & Co. during 1893 with a further twenty-six, classified Class T, between 1895 and 1898, again from Neilson & Co., during the William Pickersgill period. Both classes became Class D41 with the LNER and seen here in the scrap line at Inverurie Works is No. 2250. She had entered service during 1897 and had been withdrawn from service in December 1947.

Above: Friday, 20 August 1948. In Dundee Tay Bridge Shed yard ex-NBR Class G (LNER Class Y9) 0-4-0 saddle tank No. 68100 is looking very smart with its new British Railways identity. Entering service during 1889 from Cowlairs Works, she would be numbered 342 then 1084 by the NBR, becoming 10084 and later 8100 with the LNER. She would see seventy-one years of service before being withdrawn in 1960.

Opposite top: Friday, 20 August 1948. Four examples of the Edward Thompson development of the Nigel Gresley-designed Class V2 locomotives were constructed as Pacifics during 1944 and 1945 and designated Class A2/1. No. 60510 *Robert the Bruce*, seen here at Dundee Tay Bridge Shed, entered service from Darlington Works in January 1945. Numbered 3699 and later 510 with the LNER, she would spend most of her working life allocated either to Haymarket in Edinburgh or Ferryhill in Aberdeen, and be withdrawn in 1960 after a short working life of only fifteen years.

Opposite bottom: Friday, 20 August 1948. The William Reid design of Class J (LNER Class D30) 4-4-0s for the NBR became generally known as 'Superheated Scotts', as opposed to his earlier Class J non-superheated boiler 'Scotts'. With twenty-seven examples all coming from Cowlairs Works between 1912 and 1920, they spent much of their time working on the Waverley route and hauling Fife and Dundee expresses. No. 2430 *Jingling Geordie*, seen here in Dundee Tay Bridge Shed yard, was constructed during 1914 and would be withdrawn in 1957. Named after the well-known Edinburgh goldsmith and philanthropist George Heriot, nicknamed Jingling Geordie, who appears in the 1822 published novel by Sir Walter Scott *The Fortunes of Nigel*.

Opposite top: Sunday, 22 August 1948. At Eastfield Shed in Glasgow, ex-NBR Class F (LNER Class J88) 0-6-0 tank No. 8330 is a member of this compact-looking class of short, 11ft wheelbase, dock-shunting locomotives. Constructed at Cowlairs Works in 1905 and numbered 846 by the NBR, she would become No. 9846 and later 8330 with the LNER and be withdrawn during 1958.

Opposite bottom: Sunday, 22 August 1948. At Parkhead Shed in Glasgow wearing her new owner's identity, ex-GER Class R24 (LNER Class J69) 0-6-0 tank No. 68567 had been constructed at Stratford Works in 1896 and had been rebuilt during 1913. Transferred to the Scottish area of the LNER in 1927, she would return to English shed allocations during 1952 and be withdrawn in 1957.

Above: Sunday, 22 August 1948. Ex-NBR Class G (LNER Class Y9) 0-4-0 saddle tank No. 68117 is seen standing adjacent to the coaling tower at Kipps Shed. Constructed at Cowlairs Works in 1897 and numbered 63 by the NBR, she would become No. 9063 and later 8117 with the LNER. She would be withdrawn from service in 1962.

Sunday 22 August 1948. This spread shows three examples of the William Reid-designed Class A 0-6-2 tank for the NBR, which became Class N15 with the LNER. With a total of ninety-nine examples constructed between 1910 and 1924 by the NBL, Robert Stephenson & Co. and the North British Cowlairs Works, these sturdy, powerful and reliable locomotives handled much short-haul goods traffic around Edinburgh and Glasgow.

Above: At Kipps Shed wearing her new owner's identity, No. 69206 was constructed at Cowlairs Works in 1923 and would be withdrawn during 1960.

Opposite top: At Parkhead Shed No. 9171 has yet to be re-numbered. Constructed by the NBL during 1913, she would also be withdrawn in 1960.

Opposite bottom: At Eastfield Shed No. 9189 is in very clean condition and is carrying the No. 2 pilot disc. Constructed by the NBL in 1920, she would be withdrawn during 1958. Utilised as one of the Glasgow Queen Street Station to Cowlairs bankers giving assistance on the 1 in 42 incline, she is seen here fitted with slip-coupling equipment.

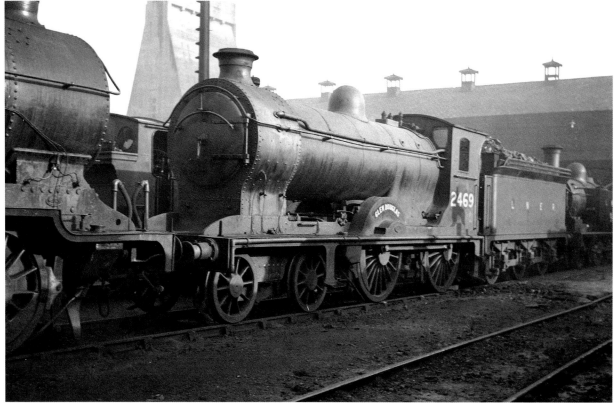

Opposite top: Sunday, 22 August 1948. Sporting green livery, ex-GNR Class H3 (LNER Class K2) 2-6-0 No. 61783 *Loch Sheil* is seen at Eastfield Shed in Glasgow. Constructed by Kitson & Co. during 1921, she would by the late 1920s be allocated to this shed to work the West Highland line and its extension to Mallaig. Given her name in 1933, she would be transferred to the GNS section in 1952 and be withdrawn from service during 1959.

Opposite bottom: Sunday, 22 August 1948. Also seen at Eastfield Shed is an example of the locomotive class that the K2s replaced on the West Highland line. Ex-NBR Class K (LNER Class D34) 4-4-0 'Glen' No. 2469 *Glen Douglas* had been constructed at Cowlairs Works during 1913 and numbered 256 by them. Becoming No. 9256 and later 2469 with the LNER, she would be withdrawn during 1959. Repainted in NBR livery later in 1959 and returned to service, she was utilised on many specials until finally withdrawn in 1962. She now resides in the Riverside Museum in Glasgow.

Above: Sunday, 22 August 1948. Bearing her new number and owner's identity, ex-LNER Class J38 0-6-0 No. 65914 is also parked in the yard at Eastfield Shed. Constructed during 1926 at Darlington Works, she would be withdrawn in 1966.

Opposite top: Sunday, 22 August 1948. The NBR Class L (LNER Class C16) 4-4-2 tanks were William Reid's superheated boiler version of his earlier Class M (LNER Class C15) saturated-boiler locomotives. Twenty-one examples were delivered between 1915 and 1921, all from the NBL in Glasgow. No. 7500, which entered service during 1921, is seen here at Eastfield Shed. Originally numbered 514 by the NBR, she would become No. 9514 and later 7500 with the LNER. She would be withdrawn from service in 1959.

Opposite bottom: Thursday, 26 August 1948. Ex-NBR Class C (LNER Class J36) 0-6-0 No. 65243 *Maude* is standing in the yard at Haymarket Shed. Constructed by Neilson & Co. during 1891 and numbered 673 by the NBR, she would be rebuilt in the form seen here in 1915. Requisitioned by the War Department in 1917, she served in France with twenty-four of her classmates and was returned to the NBR in 1919 when she was given her name after Lieutenant General Sir Frederick Stanley Maude who had served both on the Western Front and in Mesopotamia during the First World War. Withdrawn from service in 1966, she was purchased by the Scottish Railway Preservation Society and is currently on display in the museum at Bo'ness.

Above: Monday, 8 August 1949. At Kipps Shed ex-NBR Class G (LNER Class Y9) 0-4-0 saddle tank No. 68120 is sporting her new identity; note the unusual position of the number plate above the smoke-box door. She entered service from Cowlairs Works in 1899 numbered 308 by the NBR, becoming No. 9308 and later 8120 with the LNER. She would be withdrawn in 1955.

Tuesday, 9 August 1949. This wonderfully moody photograph shows one of the Edinburgh Waverley Station west-end pilot locomotives. Ex-NBR Class D (LNER Class J83) 0-6-0 tank No. 68481 is sporting a fully lined-out green livery and is in beautifully clean condition, as all the Waverley Station pilots were at this time. Constructed by Sharp, Stewart & Co. during 1901, she would be withdrawn from service in 1962.

Tuesday, 9 August 1949. At Eastfield Shed Class K1 2-6-0 No. 62023 was delivered from the NBL a few days earlier and is carrying out running-in trials. The diamond-shaped NBL works plate is clearly visible on the smoke-box side. Allocated new to Blaydon, she would spend her working life allocated to sheds in the north-east of England and would be withdrawn during 1967.

Tuesday, 9 August 1949. At the eastern end of Waverley Station in Edinburgh, ex-LNER Class A3 4-6-2 No. 60035 *Windsor Lad* is arriving at the head of a 'down' express passenger working. Entering service from Doncaster Works during 1934, she was initially numbered 2500 by the LNER and named after the racehorse that won the 1934 Derby and St Leger races. Allocated new to Haymarket Shed in Edinburgh, she would spend the bulk of her working life based there and be withdrawn from service in 1961.

Tuesday, 9 August 1949. Also seen arriving at Waverley Station with a 'down' express is ex-LNER Class A2/2 4-6-2 No. 60506 *Wolf of Badenoch*. Originally constructed at Doncaster Works during 1936 as a Class P2 2-8-2 locomotive to a design by Nigel Gresley, it became one of the six rebuilds of that class by Edward Thompson to Pacific Class A2/2. Rebuilt at Doncaster Works in 1944, she was initially allocated to Haymarket Shed in Edinburgh but during 1949 was transferred to New England Shed and would be withdrawn in 1961. 'The Wolf of Badenoch' was the nickname for the Earl of Buchan, Alexander Stewart.

Opposite top: Monday, 31 July 1950. This unnamed member of the ex-GNR Class H3 (LNER Class K2) 2-6-0 is seen at Fort William Shed. No. 61784 entered service from Kitson & Co. during 1921 and would during, the early 1920s, be allocated to Eastfield Shed in Glasgow to work on the West Highland line to Fort William. She would be withdrawn from service in 1961.

Opposite bottom: Tuesday, 1 August 1950. At Eastfield Shed ex-LNER Class D11 4-4-0 No. 62686 *The Fiery Cross* is in a filthy condition with the number and name hardly discernable. This 'Scottish Director' was constructed post-grouping as part of a batch of twelve examples entering service from Armstrong Whitworth & Co. during 1924. Initially allocated to Dundee Shed to work passenger trains to Edinburgh and Glasgow, she would end her days based at Eastfield and be withdrawn during 1961. Her name *The Fiery Cross* appears in the 1810 published poem *The Lady of the Lake* by Sir Walter Scott.

Above: Wednesday, 2 August 1950. At Parkhead Shed ex-LNER Class V1 2-6-2 tank No. 67632 is carrying a 'Bridgeton Cross' destination board. Constructed during 1935 at Doncaster Works, she would be rebuilt as a Class V3 locomotive in 1957 and be withdrawn during 1962. This design by Nigel Gresley was the only design of tank locomotives utilising this wheelbase by the LNER and found in the Scottish area working on suburban traffic around both Edinburgh and Glasgow.

Thursday, 3 August 1950. The powerful presence of this ex-LNER Class J39 0-6-0 is clearly evident with No. 4795 standing in the yard at Ferryhill Shed in Aberdeen. Constructed at Darlington Works in 1929 and allocated new to Ferryhill, she would be transferred to St Margaret's in Edinburgh and end her days based at Dunfermline Shed, being withdrawn during 1962.

Thursday, 3 August 1950. Standing on the coal ramp road at Ferryhill Shed is ex-NBR Class C (LNER Class J36) 0-6-0 No. 5213. One of the earlier members of this numerous class, she entered service from Cowlairs Works during 1889 and was rebuilt in the form seen here in 1914. After giving sixty-eight years of service, she would be withdrawn during 1957.

Thursday, 3 August 1950. Seen at Kittybrewster Shed in Aberdeen is Class B1 4-6-0 No. 61350. A product of Darlington Works in 1949, she was allocated new to Kittybrewster, but would later be transferred to St Margaret's in Edinburgh and be withdrawn during 1966, having given only seventeen years' service. This Edward Thompson design was introduced during 1942 with a total of 410 examples entering service, the last post-grouping in 1952. The NBL in Glasgow produced the greater number with 290 examples; the Vulcan Foundry constructed fifty examples and the former GCR Works at Gorton contributed ten locomotives, the remaining sixty coming out of Darlington Works.

Thursday, 3 August 1950. Ex-NER Class E1 (LNER Class J72) 0-6-0 tank No. 68719 is seen at Kittybrewster Shed. Constructed at Darlington Works during 1920, she was transferred to the Scottish area in 1932 and would be withdrawn from service in 1961. Not many classes of locomotives in Great Britain were found so useful that construction continued over a period of fifty years. This Wilson Worsdell class was introduced during 1898, with the final examples entering service during 1951. One example, No. 69023, has managed to reach the preservation scene.

Thursday, 3 August 1950. Standing adjacent to the roundhouse at Kittybrewster Shed is ex-GER Class S69 (LNER Class B12) 4-6-0 No. 61511. One of the earlier members of the class to be constructed, she entered service from Stratford Works during 1913 and was transferred to the GNS section in 1942. She had originally been constructed with a Belpaire firebox but would be rebuilt in 1946 with a round-topped variety and be withdrawn from service during 1952. Note that she has no corporate identity painted on her tender.

Thursday, 3 August 1950. Ex-GER Class M15 (LNER Class F4) 2-4-2 tank No. 67157 had recently been allocated as Inverurie Works pilot and is seen here standing in the works yard. Introduced by Thomas Worsdell during 1884, this example was constructed at Stratford Works in 1907 and was sent north in 1948 to assist in working the St Combs branch from Fraserburgh. With a very light axle weight of just less than 15 tons, she was one of only four members of the class sent to Scotland for this work. She would be the last member of this numerous class to be withdrawn in June 1956.

Friday, 4 August 1950. Sporting the experimental British Railways blue livery, ex-LNER Class A4 4-6-2 No. 60027 *Merlin* is seen here, minus coupling and connecting rods, being shunted in the yard at Haymarket Shed in Edinburgh. Constructed at Doncaster Works during 1937, she would be allocated to Haymarket before being withdrawn in 1965.

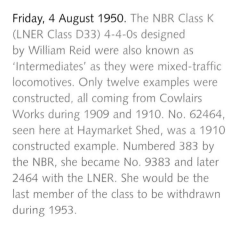

Friday, 4 August 1950. The NBR Class K (LNER Class D33) 4-4-0s designed by William Reid were also known as 'Intermediates' as they were mixed-traffic locomotives. Only twelve examples were constructed, all coming from Cowlairs Works during 1909 and 1910. No. 62464, seen here at Haymarket Shed, was a 1910 constructed example. Numbered 383 by the NBR, she became No. 9383 and later 2464 with the LNER. She would be the last member of the class to be withdrawn during 1953.

Above: Friday, 4 August 1950. Ex-NBR Class B (LNER Class J35) 0-6-0 No. 64523 is shunting some wagons at the former NBR Works at St Margaret's in Edinburgh. Constructed at Cowlairs Works in 1910 with a saturated boiler, she would be rebuilt during 1942 with a superheated boiler and be withdrawn in 1961.

Opposite top: Friday, 3 August 1951. At the former LNER Canal Shed at Carlisle, ex-NBR Class M (LNER Class C15) 4-4-2 tank No. 67474 is looking in good external condition. Constructed during 1913 by the Yorkshire Engine Co., she would be allocated to Eastfield Shed in Glasgow during 1954 and fitted with 'push & pull' equipment to work on the Craigendoran to Arrochar services. She would be one of the last two examples of the class to be withdrawn during 1960.

Opposite bottom: Saturday, 4 August 1951. Parked in Hawick Shed yard is ex-NBR Class J (LNER Class D30) 4-4-0 'Superheated Scott' No. 62423 *Dugald Dalgetty*. Constructed at Cowlairs Works in 1914, she would be withdrawn from service during 1957. The character Dugald Dalgetty appears in the 1819 published novel *A Legend of Montrose* by Sir Walter Scott.

Wednesday, 8 August 1951. Designed by Edward Thompson as the 'Standard' Pacific locomotive for the LNER, the Class A2/3 only consisted of fifteen locomotives constructed during 1946 and 1947. Seen here at Haymarket Shed, No. 60521 *Watling Street* entered service from Doncaster Works in 1947 and would be allocated new to Gateshead Shed. Named after the racehorse that won the 1942 Derby, she would end her days based at Tweedmouth Shed and be withdrawn during 1962.

Thursday, 9 August 1951. In good clean condition, ex-NER Class O (LNER Class G5) 0-4-4 tank No. 67303 is seen at Tweedmouth Shed. Entering service from Darlington Works in 1897 and numbered 1745 by the NER, she would become No. 7303 and be withdrawn from service during 1953.

August 1952. At Thornton Junction Shed, ex-NBR Class A (LNER Class N15) 0-6-2 tank No. 69153 is standing over an inspection pit. Constructed by the NBL during 1912 and numbered 916 by the NBR, she would be numbered 9916 and later 9153 by the LNER, and would be withdrawn in 1958.

August 1952. Also at Thornton Junction Shed is ex-NBR Class J (LNER Class D30) 4-4-0 'Superheated Scott' No. 62441 *Black Duncan*. The penultimate example of the class to be constructed at Cowlairs Works in 1920, she would be numbered 500 by the NBR, becoming No. 9500 and later 2441 with the LNER. She would be withdrawn during 1958.

August 1952. Ex-NBR Class C (LNER Class J36) 0-6-0 No. 65253 *Joffre* is standing adjacent to the shed at Thornton Junction. A product of Sharp, Stewart & Co. during 1892, she would spend time in France during the First World War and on her return was named in honour of Marshal Joffre who was Commander-in-Chief of the French forces on the Western Front until 1916. She would give seventy-one years of service before being withdrawn during 1963.

Monday, 11 August 1952. In sparkling condition and waiting to depart from Ballater Station with a 'stopper' to Aberdeen is ex-GER Class S69 (LNER Class B12) 4-6-0 No. 61563. Constructed at Stratford Works during 1920, she would be transferred to the GNS section of the LNER in 1939 and be withdrawn in 1953.

Opposite top: Monday, 11 August 1952. Shunting in Ballater Station yard is ex-GER Class S69 (LNER Class B12) 4-6-0 No. 61532. Constructed at Stratford Works in 1914, she would be transferred to the GNS section during 1948 and be withdrawn from service in 1953. Note the shunter precariously positioned on his shunting pole whilst the movement is made.

Opposite bottom: Tuesday, 12 August 1952. Departing from Fraserburgh Station at the head of a goods train for Aberdeen is ex-GER Class S69 (LNER Class B12) 4-6-0 No. 61539. A product of Stratford Works in 1917, she was transferred to the GNS section during 1933 and would be withdrawn in 1954.

Above: Tuesday, 12 August 1952. At Maud Station, ex-GNoSR Class F (LNER Class D40) 4-4-0 No. 62279 *Glen Grant* awaits its next duty. She was constructed by the NBL during 1920 as one of a batch of six superheated boiler examples of the class, which were ordered during the Thomas Heywood regime. She would be numbered 52 by the GNoSR, becoming No. 6852 and later 2279 with the LNER. Withdrawal would come in 1955. The locomotive name was that of the residence of Mr James Grant, a director of the company.

Wednesday, 13 August 1952. In Kittybrewster Shed yard ex-NER Class E1 (LNER Class J72) 0-6-0 tank No. 68700 is carrying two lamps and a shunter's pole on its front buffer beam in preparation for yard pilot duty. Entering service from Darlington Works during 1914, she would be transferred to Scotland in 1932 and withdrawn from service during 1958.

Wednesday, 13 August 1952. Standing beside the coaling tower at Kittybrewster Shed in beautifully clean condition is ex-GNoSR Class F (LNER Class D40) 4-4-0 No. 62276 *Andrew Bain*. Another example of a batch of six members of the class constructed by the NBL, she entered service during 1920 and was numbered 48, becoming No. 6848 and later 2276 with the LNER. She would be withdrawn from service in 1955.

This close-up of the nameplate on No. 62276 *Andrew Bain* also shows the detail on the works plate, 'London & North Eastern Railway N.B.Loco. Co. 1920'. Mr Andrew Bain had been vice chairman and chairman of the company.

Wednesday, 13 August 1952. During 1915 the GNoSR had two locomotives delivered from Manning Wardle & Co. to work the harbour trust lines in Aberdeen. They were classified Class X with them, later becoming Class Z4 with the LNER. Seen here working on the Aberdeen harbour lines is the second of the pair, No. 68191, which had initially been numbered 117, becoming No. 44 with the GNoSR and later 6844 and 8191 with the LNER. She would be withdrawn during 1959.

Wednesday, 13 August 1952. Being squeezed onto the turntable at Elgin Shed is Class B1 4-6-0 No. 61323 bearing a 61A Kittybrewster shed code. Constructed during 1948 by the NBL, she would be withdrawn after a relatively short career of only fifteen years in 1963.

Wednesday, 13 August 1952. Also seen in Elgin Shed yard whilst taking water is ex-GNoSR Class T (LNER Class D41) 4-4-0 No. 62241. Entering service from Neilson & Co. in 1895 during the William Pickersgill period, she would be numbered 97 by the GNoSR, becoming 6897 and later 2241 with the LNER. She would give fifty-eight years of service before being withdrawn in 1953.

Above: Wednesday, 13 August 1952. Bearing a 61A Kittybrewster shed code, Class B1 4-6-0 No. 61401 is waiting to depart from Aberdeen Station. One of the last examples of the class to enter service from Darlington Works during 1950, she would only give fourteen years of service before being withdrawn in 1964.

Opposite: Friday, 15 August 1952. At Craigellachie Station the graceful lines of this ex-GNoSR Class T (LNER Class D41) 4-4-0 No. 62248 are clearly evident as she waits to depart with a working to Boat of Garten via the Speyside line. Constructed by Neilson & Co. in 1897, she would be withdrawn from service two months after this photograph, in October 1952.

Sunday, 30 August 1953. Ex-NBR Class C (LNER Class J36) 0-6-0 No. 65235 *Gough* is seen standing at Haymarket Shed. An example of the class constructed at Cowlairs Works in 1891, she would see service in France and on return be named after General Gough, who was commander of the British Fifth Army between 1916 and 1918. She would give seventy years' service before being withdrawn during 1961.

Sunday, 30 August 1953. Seen at St Margaret's Shed in Edinburgh is resident ex-NBR Class G (LNER Class Y9) 0-4-0 saddle tank No. 68098. Constructed at Cowlairs Works during 1887, she would give sixty-seven years of service before being withdrawn in 1954.

Tuesday, 1 September 1953. At Eastfield Shed stands the impressive bulk of ex-LNER Class Q1 0-8-0 No. 69925, which was being utilised on shunting duties at Cadder yard in Glasgow. Originally constructed by Kitson & Co. for the GCR as part of their Class 8A (LNER Class Q4) 0-8-0 tender locomotives, she was rebuilt in the Edward Thompson era during 1942 as an 0-8-0 tank locomotive along with a further twelve classmates. She would be withdrawn during 1954.

Tuesday, 1 September 1953. At Polmont Shed dock-shunting tank ex-NBR Class F (LNER Class J88) 0-6-0 tank No. 68354 looks to be in ex-works condition. Constructed at Cowlairs Works in 1919, she was numbered 289 by the NBR, later becoming No. 9289 and 8354 with the LNER. She would be withdrawn in 1960.

Above: Thursday, 3 September 1953. The lines between the Mound and Haymarket tunnels heading west out of Waverley Station in Edinburgh cut through Princes Street Gardens which became a favourite site for enthusiasts to gather and photograph train movements. On this day ex-NBR Class B (LNER Class J37) No. 64555 0-6-0 is working an eastbound unfitted goods train. Entering service from Cowlairs Works in 1916, she would be withdrawn after forty-eight years' service during 1964. Note the S&T man ducking for cover as the train approaches.

Opposite: Thursday, 3 September 1953. Proceeding through Princes Street Gardens cutting at the head of a Fife-bound 'stopper' is ex-LNER Class D49 4-4-0 No. 62708 *Argyllshire*. Constructed at Darlington Works during 1928, she would be withdrawn in 1959.

Thursday, 3 September 1953. Heading through Princes Street Gardens cutting 'light engine' is ex-NBR Class K (LNER Class D34) 4-4-0 'Glen' No. 62495 *Glen Luss.* Constructed at Cowlairs Works during 1920, she would be withdrawn in 1961.

Thursday, 3 September 1953. Ex-NBR Class C (LNER Class J36) 0-6-0 No. 65216 *Byng* is seen in the yard of former NBR Shed at Carlisle Canal. Entering service during 1890 from Cowlairs Works and rebuilt in the form seen here in 1915, she would acquire her name after service in France. Field Marshal Byng was the commander of the Canadian Corps at the Battle of Vimy Ridge. She would be withdrawn in 1962.

Tuesday, 24 August 1954. At Dundee Tay Bridge Shed yard, ex-NBR Class C (LNER Class J36) 0-6-0 No. 65217 *French* has its tender fitted with a weather cab. Constructed at Cowlairs Works during 1890, she would be rebuilt in 1913 in the form seen here and be requisitioned by the Railway Operating Division (ROD) during 1917. Serving in France, she would be returned in 1919 and consequently named after Field Marshal French who was Commander-in-Chief of the British Expeditionary Force at the beginning of the First World War. She would be withdrawn during 1962 after seventy-two years of service.

Thursday, 26 August 1954. In ex-works condition at Inverurie Works, ex-LNER Class D11 4-4-0 'Scottish Director' No. 62671 *Bailie MacWheeble* is waiting to return to service. The first of a batch of twelve examples constructed post-grouping by Kitson & Co. of the former GCR Class 11F in 1924, she would be withdrawn during 1961. The character Bailie MacWheeble appears in the 1814 published novel *Waverley* by Sir Walter Scott.

Above: Sunday, 29 August 1954. Carrying the correct 64A shed code and standing on one of the former roundhouse turntable roads at St Margaret's Shed is ex-NBR Class A (LNER Class N15) 0-6-2 tank No. 69152 in a very clean condition. Entering service from the NBL during 1912, she would be numbered 915 by the NBR, becoming No. 9915 and later 9152 with the LNER. She would be withdrawn in 1958.

Opposite top: Monday, 30 August 1954. Manoeuvring in Bathgate Shed yard is ex-NBR Class C (LNER Class J36) 0-6-0 No. 65248. Constructed at Cowlairs Works in 1891, she would be rebuilt in the form seen here during 1915 and be withdrawn from service in 1956.

Opposite bottom: Monday, 30 August 1954. Also seen in Bathgate Shed yard is ex-NBR Class B (LNER Class J35) 0-6-0 No. 64504. Entering service in 1910 from the NBL and constructed with a saturated boiler, she would be rebuilt during 1933 with a superheating boiler and be withdrawn after fifty years' service in 1960.

Above: Wednesday, 1 September 1954. Standing in the yard at Kipps Shed is ex-NBR Class C (LNER Class J36) 0-6-0 No. 65260. Constructed as part of a batch of fifteen examples from Sharp, Stewart & Co., which entered service during 1892, she would be rebuilt in the form seen here in 1915. Numbered 689 by the NBR, she would become No. 9689 and later 5260 with the LNER and be withdrawn from service during 1962 having given seventy years' service.

Opposite top: Wednesday, 1 September 1954. Ex-GNR Class N2 (LNER Class N2) 0-6-2 tank No. 69518 is seen here parked in Kipps Shed yard. She was a product of the NBL during 1921, the class designed to work suburban traffic out of London King's Cross Station. Transferred to Glasgow during 1927 to work suburban traffic around the Glasgow area, she would be withdrawn from service in 1961.

Opposite bottom: Thursday, 2 September 1954. At Polmont Shed, ex-NBR Class C (LNER Class J36) 0-6-0 No. 65233 *Plumer* is simmering gently. Constructed at Cowlairs Works in 1891 and rebuilt in the form seen here during 1913, she would serve in France during the First World War. On return she would be named after Field Marshal Plumer, who had commanded V Corps at the Second Battle of Ypres. She would give sixty-nine years of service before being withdrawn during 1960.

Opposite: Monday, 25 July 1955. At the head of an 'up' express, ex-LNER Class V2 2-6-2 No. 60819 is seen accelerating away from Montrose Station and about to cross the single-track bridge spanning the River South Esk. Bearing a 61B Ferryhill shed code, she had entered service from Darlington Works during 1937 and would be withdrawn in 1962.

Above: Monday, 25 July 1955. Waiting to depart from Waverley Station in Edinburgh with the 2.15 p.m. working to Aberdeen is ex-LNER Class A2/1 4-6-2 No. 60509 *Waverley*. A further example of the Edward Thompson development of the Class V2 locomotives, she entered service from Darlington Works during 1945 and was allocated new to Haymarket for much of her working life. Named after the principal character in the novel *Waverley* by Sir Walter Scott, which was published in 1814, she would be withdrawn in 1960.

Opposite: Tuesday, 26 July 1955. Seen here taking water whilst standing over an ash pit in the yard at Kittybrewster Shed is BR Standard Class 4 2-6-4 tank No. 80107. A product of Doncaster Works during 1954, she was initially allocated to Polmadie Shed in Glasgow but soon found her way to Kittybrewster. She would be withdrawn from service during 1964.

Above: Tuesday, 26 July 1955. In Kittybrewster Shed yard ex-NBR Class K (LNER Class D34) 4-4-0 'Glen' No. 62489 *Glen Dessary* awaits her next duty. Constructed at Cowlairs Works during 1920, she would be numbered 490 by the NBR, becoming No. 9490 and later 2489 with the LNER. She would be withdrawn from service in 1959.

Left: Tuesday, 26 July 1955. Carrying 'The Granite City' headboard at Ferryhill Shed is BR Standard Class 5 4-6-0 No. 73007. Bearing a 63A Perth shed code, she had been allocated new there after construction at Derby Works during 1951. She would end her days based at Stirling Shed, being withdrawn after only sixteen years of service in 1966.

Below: Friday, 29 July 1955. Seen here on the turntable at Dundee West Shed is Class 5 4-6-0 No. 44677. The fireman is busy trimming coal in the self-weighing tender whilst the driver is operating the vacuum-driven turntable. Constructed at Horwich Works during 1950, she would be withdrawn in 1967.

Friday, 29 July 1955. With a protective weatherboard fitted to her tender, ex-NBR Class C (LNER Class J36) 0-6-0 No. 65345 is seen in Thornton Junction Shed yard. Entering service during 1900 as the penultimate example of the class constructed at Cowlairs Works, she would be rebuilt in the form seen here in 1923 and be destined to be one of the last pair of J36s to be withdrawn from service in June 1967.

Friday, 29 July 1955. At Thornton Junction Shed ex-GER Class R24 (LNER Class J69) 0-6-0 tank No. 68504, which had been constructed at Stratford Works in 1890, is looking rather tired and careworn. Transferred to the Scottish area during 1928, she would be withdrawn after sixty-six years of service in 1956.

Friday, 29 July 1955. Also seen at Thornton is ex-LNER Class D11 4-4-0 'Scottish Director' No. 62681 *Captain Craigengelt*. Constructed post-grouping by Kitson & Co. during 1924, this example of the former GCR Class 11F would be named after a character who appears in the Sir Walter Scott novel *The Bride of Lammermuir* published in 1819. She would be withdrawn during 1961.

Sunday, 31 July 1955. Standing on one of the turntable roads at St Margaret's Shed in Edinburgh is ex-NBR Class D (LNER Class J83) 0-6-0 tank No. 68474. Constructed by Sharp, Stewart & Co. in 1901, she would be numbered 827 by the NBR, becoming No. 9827 and later 8474 with the LNER. Utilised over many years as one of the Edinburgh Waverley Station pilots, she would be withdrawn during 1958.

Opposite: Monday, 1 August 1955. At Stirling Station ex-NBR Class J (LNER Class D30) 4-4-0 'Superheated Scott' No. 62426 *Cuddie Headrigg* is waiting to depart with a 'stopper' for the Fife line. Constructed at Cowlairs Works during 1914 and numbered 417 by the NBR, she would become No. 9417 and later 2426 with the LNER. Named after the servant character who appears in the 1816 published novel *Old Mortality* by Sir Walter Scott, she would be withdrawn in 1960.

Above: Tuesday, 2 August 1955. This close-up of the name painted on the front driver splasher of No. 62426 also shows the works plate: 'London & North Eastern Railway 2426 Cowlairs Works 1914'.

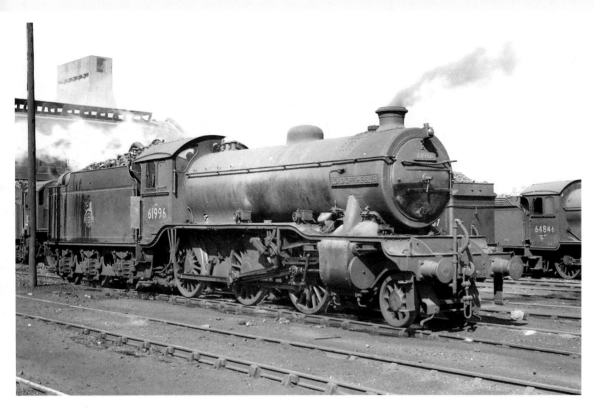

Tuesday, 2 August 1955. The powerful three-cylinder LNER Class K4 2-6-0s were designed specifically to work the arduous West Highland line and its extension to Mallaig; the first of only six members of the class entered service during 1937, followed in 1938 and 1939 by the remaining five locomotives, all from Darlington Works. All were initially allocated to Eastfield Shed in Glasgow. No. 61996 *Lord of the Isles* is seen here at Eastfield Shed. Entering service during 1938 and numbered 3444 by the LNER, she would be withdrawn in 1961.

Wednesday, 3 August 1955. At Parkhead Shed, ex-LNER Class V1 2-6-2 tank No. 67613 is seen carrying a Bridgeton Central headboard. Constructed at Doncaster Works in 1931, she would be rebuilt as a Class V3 locomotive during 1956 and be withdrawn in 1962.

Wednesday, 3 August 1955. At Motherwell Shed, resident ex-WD Class 8F 2-10-0 No. 90756 is looking in particularly clean condition. Constructed during 1945 by the NBL and numbered 73780 by the WD, she would never see service abroad but became British Railways property in 1948. Allocated to Motherwell for her entire BR lifetime, she would be withdrawn in 1962.

Thursday, 4 August 1955. Dugald Drummond introduced the Class 294 0-6-0 locomotives to the Caledonian Railway in 1883, with construction continuing until 1897. Classified 2F by the LMS, No. 57309 is seen here at Corkerhill Shed attached to the breakdown train. Constructed at St Rollox Works during 1887, she would give seventy-six years of service, being withdrawn in 1963.

Thursday, 4 August 1955. A later and more powerful class of 0-6-0s introduced to the Caley in 1899 and designed by John McIntosh were the Class 812 locomotives. Seen here at Dawsholm Shed is No. 57652, constructed at St Rollox Works in 1899 and numbered 824 by the CR, she would become No. 17652 with the LMS and be withdrawn during 1962. Classified 3F by the LMS, examples were constructed at Dübs & Co., Neilson Reid & Co. and Sharp, Stewart & Co., in addition to the Caley's own St Rollox Works.

Thursday, 4 August 1955. Seen near Corkerhill in Glasgow is ex-LMS Class 5MT 2-6-0 No. 42909 at the head of a train of loaded mineral wagons. Constructed at Crewe Works during 1930, she would be withdrawn in 1966.

This George Hughes design for the LMS was introduced during 1926, with a total of 245 examples being constructed at both Crewe and Horwich Works.

Friday, 5 August 1955. With a group of spectators watching 'The Elizabethan' depart from platform 10 at Waverley Station in Edinburgh, ex-LNER Class A4 4-6-2 No. 60034 *Lord Faringdon* is in charge. The last of the class to enter service from Doncaster Works in 1938, she was initially named *Peregrine* but was renamed in 1948 after the 1st Baron Faringdon, Alexander Hamilton, who had been chairman of the Great Central Railway from 1899 until 1922. Allocated to King's Cross Shed for most of her working life, she would end her days based at Ferryhill in Aberdeen before being withdrawn during 1966.

Tuesday, 14 August 1956. At the former LNER Canal Shed in Carlisle, ex-LNER Class A3 4-6-2 No. 60041 *Salmon Trout* is standing in the yard. Constructed at Doncaster Works during 1934, she would be named after the racehorse that won the 1924 St Leger. Allocated new to Haymarket in Edinburgh, she would move to St Margaret's Shed in 1960 and be withdrawn during 1965.

Tuesday, 14 August 1956. Also seen at Canal Shed in Carlisle is ex-NBR Class J (LNER Class D30) 4-4-0 'Superheated Scott' No. 62432 *Quentin Durward*. Entering service from Cowlairs Works in 1914, she would be withdrawn during 1958. She was named after the principal character in the 1823 published Sir Walter Scott novel *Quentin Durward*.

Tuesday, 21 August 1956. Seen passing Hurlford Shed at the head of a goods train is ex-CR Class 294 (LMS Class 2F) 0-6-0 No. 57295 carrying pilot disc No. K111. Constructed at St Rollox Works during 1887 to a design by Dugald Drummond, she would end her days based at Hurlford and be withdrawn after seventy-five years of service in 1962.

Tuesday, 30 July 1957. Entering service during 1948 and allocated to Ferryhill Shed in Aberdeen in 1949, Class A2 4-6-2 No. 60531 *Bahram*, seen here at her home shed, was transferred to York at the end of 1962 from where she was withdrawn. Constructed at Doncaster Works, she was named after the racehorse that won three of the Classics during 1935: the Derby, the 2,000 Guineas and the St Leger.

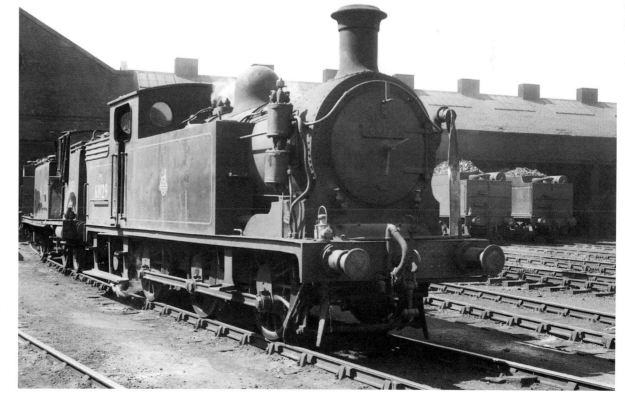

Tuesday, 30 July 1957. Also seen at Ferryhill Shed, bathed in strong sunlight, is ex-NBR Class A (LNER Class N15) 0-6-2 tank No. 69129. Constructed by the NBL in 1910, she would be based at Ferryhill during her British Railway days, being withdrawn during 1958.

Tuesday, 30 July 1957. Purchased from Manning Wardle & Co. during 1915, the two Class Y (LNER Class Z5) 0-4-2 tanks were slightly heavier than the later two purchases from the same company, Class X (LNER Class Z4).

Utilised for dock shunting, Class Z5 No. 68192 is seen here at Kittybrewster Shed. She would be withdrawn during 1960.

Wednesday, 31 July 1957.
At Aberfeldy Station, milk churns are being loaded onto the train before it departs for the branch junction at Ballinluig. Ex-CR Class 439 (LMS Class 2P) 0-4-4 tank No. 55226 had been constructed at St Rollox Works in 1914 and would be withdrawn during 1961.

Thursday, 1 August 1957.
At Thornton Junction Shed yard, ex-NBR Class J (LNER Class D30) 4-4-0 'Superheated Scott' No. 62442 *Simon Glover* awaits her next duty. Constructed at Cowlairs Works during 1920 as the last example of the class of twenty-seven locomotives, she would be numbered 501 by the NBR, becoming No. 9501 and later 2442 with the LNER, and be withdrawn in 1958. She was named after a principal character in the Sir Walter Scott novel *The Fair Maid of Perth*, published in 1828.

Friday, 2 August 1957. Looking to be in ex-works condition, ex-CR Class 439 (LMS Class 2P) 0-4-4 tank No. 55238 is bearing a 65F Grangemouth shed code and is seen in the yard there. Constructed as part of a batch of four examples of the class by St Rollox Works in 1922 during the William Pickersgill period, she would be withdrawn in 1961.

Friday, 2 August 1957. Ex-NBR Class J (LNER Class D30) 4-4-0 'Superheated Scott' No. 62426 *Cuddie Headrigg* is seen here again, this time at Dalry Road Shed in Edinburgh. During this period she was based at Stirling Shed to work the route into Princes Street Station in Edinburgh.

Friday, 2 August 1957. At the head of an Aberdeen-bound express from Edinburgh, Class A1 4-6-2 No. 60159 *Bonnie Dundee* has just burst out of the Mound Tunnel into the cutting at Princes Street Gardens. Constructed at Doncaster Works during 1949 and allocated new to Haymarket Shed in Edinburgh, she would be based there for her entire short working life, being withdrawn after only fourteen years' service in 1963. The 7th Laird of Claverhouse, also known as Bonnie Dundee, is the subject of the poem written by Sir Walter Scott published in 1825.

Friday, 2 August 1957. Another Aberdeen express sees ex-LNER Class V2 2-6-2 No. 60873 *Coldstreamer* bursting from the Mound Tunnel into the cutting at Princes Street Gardens. Constructed at Doncaster Works in 1939, she would end her days based at St Margaret's Shed in Edinburgh and be withdrawn during 1962.

Sunday, 4 August 1957. At Eastfield Shed in Glasgow ex-GNR Class H3 (LNER Class K2) 2-6-0 No. 61776 is parked outside the shed. Entering service from Kitson & Co. during 1921, she would be transferred to St Margaret's Shed in Edinburgh during 1932 and by 1942 was based at Eastfield. She would be withdrawn from service in 1959.

Sunday, 4 August 1957. Standing in St Rollox Shed yard is BR Standard Class 5 4-6-0 No. 73147. Constructed at Derby Works in 1957, she was an example of the thirty locomotives of the class fitted with Caprotti British valve gear. Allocated new to St Rollox, she would be withdrawn after only eight years of service during 1965.

Tuesday, 6 August 1957. Seen in Hamilton Shed yard is BR Standard Class 3 2-6-0 No. 77006. Constructed at Swindon Works in 1954 and allocated new to Hamilton, she would end her days based at Motherwell Shed and be withdrawn during 1966.

Tuesday, 6 August 1957. This splendid photograph shows ex-LMS Class 5 4-6-0 No. 45433 departing from Glasgow Central Station at the head of a 'stopper' to Hamilton, which is indicated by the former Caledonian Railway semaphore route indicators still in use at this time. Constructed by Armstrong Whitworth & Co. during 1937, she was initially allocated to English sheds but was transferred to Motherwell in 1951 and would end her days based there, being withdrawn during 1966.

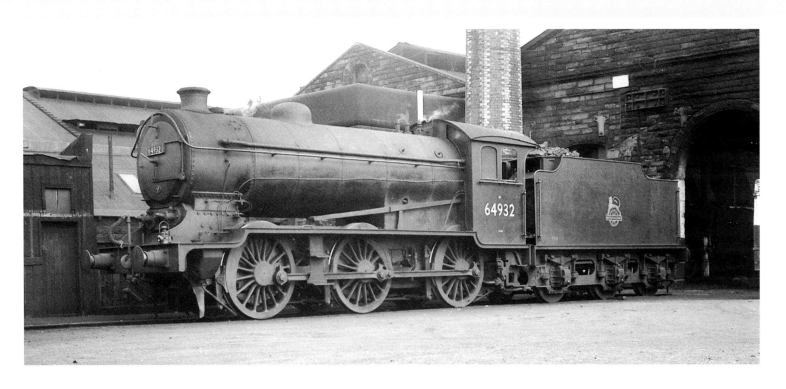

Opposite top: Wednesday, 7 August 1957. Standing adjacent to the former LNER Carlisle Canal Shed is ex-LNER Class J39 0-6-0 No. 64932. She was the last example of a batch of twenty-eight class members constructed by Beyer, Peacock & Co. in 1937. She would be withdrawn from service during 1961.

Opposite bottom: Sunday, 24 August 1958. Parked in the yard at Parkhead Shed is resident ex-GNR Class N2 (LNER Class N2) 0-6-2 tank No. 69507. Constructed by the NBL during 1920, she would give forty years of service, being withdrawn in 1960.

Above: Sunday, 24 August 1958. In good clean external condition, ex-CR Class 439 (LMS Class 2P) 0-4-4 tank No. 55224 is standing in the yard at its home shed, 66A Polmadie. Entering service from St Rollox Works during 1914, she would be withdrawn in 1961.

Sunday, 24 August 1958. Ideally suited to tight-radius curves, the John McIntosh-designed Class 498 0-6-0 tanks for the Caledonian Railway incorporated a very short wheelbase of only 10ft. A total of twenty-three examples were constructed at St Rollox Works between 1912 and 1921, and were described as 'Dock' tanks and classified 2F by the LMS. No. 56152 entered service in 1912 and would be numbered 499 by the Caley, becoming No. 16152 with the LMS. Seen here at Polmadie Shed shunting wagons, she would be withdrawn during 1959.

Sunday, 26 August 1962. Designed to replace the many ageing tank locomotives handling suburban passenger traffic, the British Railways Standard Class 4 2-6-4 tank was based on the former LMS Class 4P. Designed at the former Southern Railway Works at Brighton, the first examples entered service with the Southern Region during 1951 and a total of 155 were constructed: 130 at Brighton Works, 15 at Derby Works and 10 at Doncaster Works. They were allocated throughout the British Railway regions, except for the Western Region, which already had a sufficiency of large 2-6-2 tank locomotives handling their suburban passenger traffic. No. 80061 is seen here at Dalry Road Shed in Edinburgh bearing a 65J Balloch shed code. Constructed at Brighton Works in 1953, she would initially be allocated to Bedford Shed but by 1964 was at Dumfries and in 1965 was working out of Polmadie in Glasgow. She would be withdrawn during 1966.

This spread shows three ex-LMS Class 5 locomotives, all constructed in different decades throughout the life of the LMS Railway.

Opposite top: Tuesday, 28 August 1962. From the 1940s and seen here at Perth Shed is 'Black 5' No. 44976. Constructed at Crewe Works in 1946, she would be withdrawn during 1964.

Opposite bottom: Tuesday, 28 August 1962. Seen at Eastfield Shed and from the 1920s is 'Crab' No. 42757, which was also constructed at Crewe Works. Entering service during 1927, she would be withdrawn in 1964.

Above: Saturday, 20 July 1963. At Stirling Station, 'Black 5' No. 45053 is working an 'up' passenger train. One of the earliest members of the class constructed by the Vulcan Foundry in 1934, she would give thirty-two years of service before being withdrawn during 1966.

Wednesday, 17 June 1964. At Thornton Junction Shed on this day are seen three versions of the 0-6-0 goods locomotives utilised on traffic coming out of the Fife coalfields.

Opposite top: Ex-NBR Class C (LNER Class J36) No. 65345 has its tender fitted with a weather-protection cab. Constructed at Cowlairs Works during 1900, she would be withdrawn in 1967.

Opposite bottom: Ex-NBR Class B (LNER Class J37) No. 64618 entered service from the NBL in 1920 and would be withdrawn during 1966.

Above: Ex-LNER Class J38 No. 65925 was one of a class of thirty-five locomotives specifically designed to handle the coal traffic in Scotland. She had been constructed at Darlington Works in 1926 and would be withdrawn during 1966.

Wednesday, 17 June 1964. Seen working a goods train near Thornton Junction Shed with the buildings of the Rothes Colliery in the background is Class B1 4-6-0 No. 61292. Entering service post-nationalisation from the NBL in early 1948, she would be withdrawn from service during 1965.

Wednesday, 17 June 1964. Standing in Dunfermline Shed yard is ex-NBR Class S (LNER Class J37) 0-6-0 No. 64599. Constructed by the NBL during 1919, she would be withdrawn in 1965.

Thursday, 18 June 1964. Seen passing Saughton Junction, west of Edinburgh, BR Standard Class 4 2-6-4 tank No. 80063 is working an 'up' train from Stirling. Constructed at Brighton Works in 1953, she would only give thirteen years of service before being withdrawn during 1966. Clearly Edinburgh housewives were not concerned about smoke emissions.

Above: Friday, 13 August 1965. Standing at the former Esplanade Station at Dundee is Class 2MT 2-6-0 No. 46463. Constructed post-nationalisation at Crewe Works in 1950, she would be allocated new to Dundee Tay Bridge Shed and is carrying the correct 62B shed code. She would be withdrawn from service during 1966.

Opposite top: Friday, 13 August 1965. In a disgraceful external state, indicative of the period, BR Standard Class 5 4-6-0 No. 73153 is seen on the turntable at Perth Shed. She is an example of the thirty locomotives constructed at Derby Works during 1956 and 1957 that were fitted with British Caprotti valve gear. Allocated new to St Rollox Shed in Glasgow in 1957, she would be withdrawn from service during 1966 having given only nine short years' service.

Opposite bottom: Friday, 13 August 1965. At Dundee Tay Bridge Shed, ex-LNER Class V2 2-6-2 No. 60844 is awaiting her next duty. Constructed at Darlington Works during 1939, she only had three further months of service before being withdrawn in November 1965.

The final developments of the Class A2 locomotives under the stewardship of Arthur Peppercorn saw the construction of fifteen examples all coming from Doncaster Works, commencing with No. 525 in December 1947 and culminating with No. 60539 during August 1948. Eleven members of the class quickly found their way to Scottish sheds, being allocated to Ferryhill in Aberdeen, Haymarket in Edinburgh and Dundee Tay Bridge to handle the express passenger traffic between those cities. No. 60530 *Sayajirao* entered service during March 1948 and was named after the racehorse that won the 1947 St Leger.

Opposite: Friday, 13 August 1965. Seen here at Dundee Tay Bridge Shed is a close-up of the front driver's side of the locomotive, clearly showing the detail of the cylinder block and the nameplate.

Above: Tuesday, 23 August 1966. Just over a year later, she is seen again at Dundee Tay Bridge Shed three months before being withdrawn during the November of that year.

Opposite top: Wednesday, 24 August 1966. Standing adjacent to the shed at Dundee Tay Bridge is ex-LNER Class V2 2-6-2 No. 60813. Constructed at Darlington Works during 1937, she would be numbered 4874 by the LNER, later becoming No. 813. Fitted with the distinctive stovepipe chimney and small deflectors in 1946, she would be withdrawn later in 1966.

Opposite bottom: Wednesday, 24 August 1966. Also seen in Dundee Tay Bridge Shed yard is ex-LNER Class B1 4-6-0 No. 61102. Constructed by the NBL in 1946, she would be withdrawn from service during 1967.

Above: Friday, 26 August 1966. Standing adjacent to the coaling stage at Montrose Shed is ex-NBR Class B (LNER Class J37) 0-6-0 No. 64576 bearing a 62B Dundee Shed code. Constructed by the NBL in 1918, she would be withdrawn during 1967.

Friday, 26 August 1966. Photographed at Aberdeen Station whilst waiting to depart at the head of a Glasgow-bound service, the fireman's side nameplate on ex-LNER Class A4 4-6-2 No. 60024 *Kingfisher* appears to be cleaned to the usual Ferryhill standard. Constructed at Doncaster Works during 1936 and numbered 4483 by the LNER, she would spend time allocated to Haymarket, Doncaster, King's Cross and St Margaret's sheds, before ending her days at Ferryhill and being withdrawn from service later in 1966.